KUMAYIKOS
Ascencio Methodology
For Odd Order Magic Squares

First book in history to offer a method por getting magic squares trough multiplication

José Ascencio

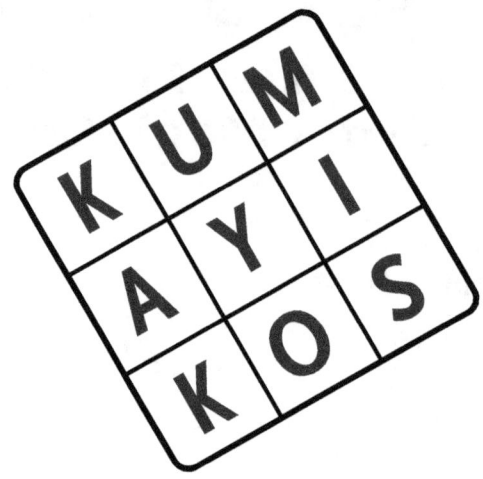

All rights reserved. The total or partial reproduction of this work is not allowed, nor its incorporation into a computer system, nor its transmission in any form or by any means (electronic, mechanical, photocopying, recording or others) without the prior written authorization of the owner of the copyright. The infringement of said rights may constitute a crime against intellectual property.

The content of this work is the responsibility of the author. All images, formulas and methods described here are the intellectual property of the author, who is solely responsible for their rights.

Book layout and Cover Design: José Ascencio Millán

Copyright 2022: José Ascencio Millán

ISBN paperback: 9798809917872

To my Father God, and his son Jesus, creators of the entire universe and the language with which we can understand it, mathematics.

To Raquel my wife, who when she saw me falter in the completion of this book, urged me over and over again not to leave it unfinished, encouraging me to publish it, no one but she lived with me throughout this long and exciting process, I love her.

To my son and daughter Karla Guadalupe and José Diego, my beloved entrepreneurs.

To my mother Julia, who is no longer with me, I know she would have liked to leaf through and read this book, she is the first person I have in my memory holding an open book in her hands, reading to my sisters Elvia and Leticia, to whom I dedicate also this publication together with the rest of my sisters Gabriela, Rosario and Jenny.

To my friend and colleague, Roberto Santos, a man of integrity and loyalty like few others, whom I greatly admire and esteem.

Now I only ask God for the opportunity that the wise man who taught me to build my first magic square more than 30 years ago on a napkin, can see this book published, that strong man, whom I love, respect and admire with all my soul and to whom I especially dedicate this work for having inspired it, is the man who, together with my mother, gave me life

JOSE ASCENCIO ALVAREZ

Table of Contents

Introduction 1

1. Definitions 3
2. Mathematical curiosities 11
3. De la Loubère Method 19
4. Central Position Method 25
5. 180 degree pairwise method 29
6. Method for solving the Lu-Shu theorem 35
7. Multiplication Method 41
8. Exercises to reinforce 53
9. Solutions 63

KUMAYIKOS

¿What will you learn in this book?

To generate and solve odd magic squares through revolutionary methods, as well as the first method in history to do it *through multiplication*, through simple steps based on logarithmic laws, regardless of the number or size of the square, you can generate squares with the constant magic that you choose, through addition or multiplication.

> The only and first book on the planet to offer a method for obtain magic squares through MULTIPLICATIONS

8	1	6
3	5	7
4	9	2

Let me introduce myself

I am **José Ascencio Millán**, proudly Mexican and intellectual creator of this exciting and original methodology to generate magic squares, which I baptize with my father's last name, for being the one who taught me to make my first magic square on a napkin, a few decades ago, and once I had improved it, my only wish was to share it with the world.

First part...

¡ A little bit of history !

Introduction

CHINA: The story or legend tells that 2,200 years BC, a Chinese emperor named Yu, while walking on the banks of the yellow river, observed a turtle with some inscriptions on its shell, the emperor observed them carefully and realized that it was a square with 9 cells containing the values from 1 to 9, which when added by columns, rows and diagonals always gave the same result, 15, he immediately had it copied on clay tablets and since then magical, astral and medicinal properties were attributed to it.

FRANCE: Around 1691, a French ambassador named La Loubère published in the second volume of his book "Du Royaume de Siam" (The Kingdom of Siam) the first valid and effective method for the elaboration of odd order magic squares by means of the operator arithmetic of addition, the method itself was not invented by him, but he learned it from a French doctor named M. Vincent, on one of his boat trips, who in turn had learned it in the city of Surat, in India, so we could say that the method itself was developed in India, La Loubère's credit lies merely in having published it.

MEXICO: 330 years after the publication of the Loubère method, the book you are holding in your hands, KUMAYIKOS, is published, the first book in history that offers a completely new method to obtain magic squares of odd order through multiplication , as well as a completely new methodology to elaborate magic squares through different techniques that allow you to appreciate the perfect symmetry that this type of square hides in its composition, characteristics that are not appreciated with the naked eye or are not perceived through the method of La Loubere.

The methods you will learn throughout this book are:

1. de La Loubère method
2. Central position method ^(Ascencio methodology)
3. 180 degree pairwise balancing method ^(Ascencio methodology)
4. Lu-Shu teorem method ^(Ascencio methodology)
5. Multiplication method, ^(Ascencio methodology)

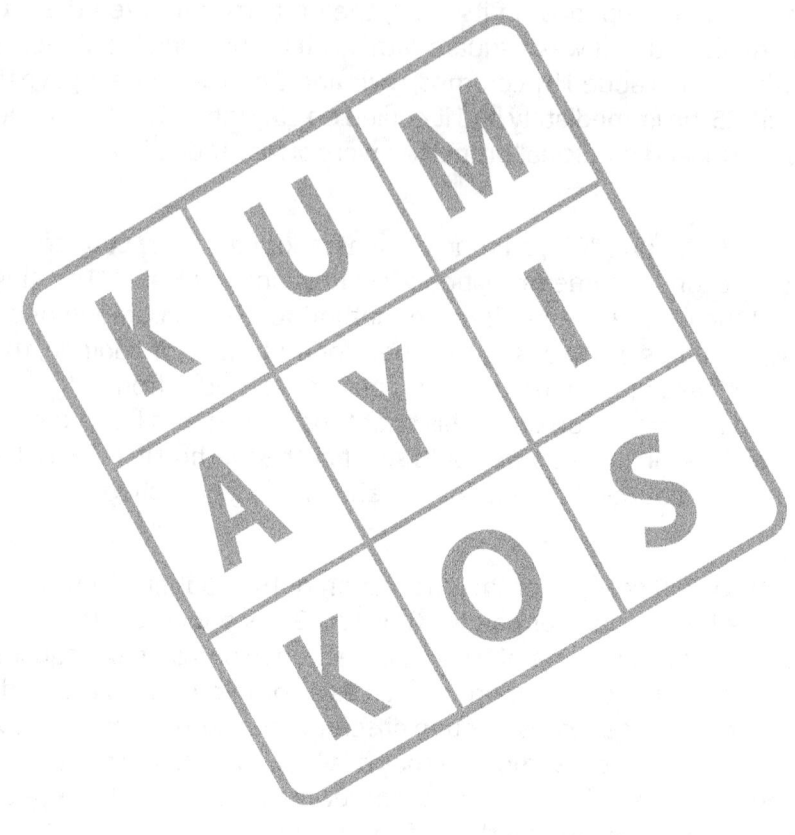

1. Definitions

first thing's first!

What is a magic square? Let us see, according to Wikipedia, "is called a magic square if the sums of the numbers in each row, each column, and both main diagonals are the same. The order of the magic square is the number of integers along one side (n), and the constant sum is called the magic constant. If the array includes just the positive integers the magic square is said to be normal. Some authors take magic square to mean normal magic square".

Note that in the definition itself it is emphasized that the result obtained is by addition and that the numbers used must be whole and consecutive numbers. That rule applies before the appearance of this book, with Kumayikos the result can be obtained through multiplication, the numbers do not necessarily have to be consecutive and much less only apply to integers, a Kumayiko can be generated with both integers, positive, negative, as well as with decimals.

Figure 1 shows the original magic square, the best known in which the sum of its values or constant always gives 15 as a result.,

FIGURE 1

8	1	6
3	5	7
4	9	2

Before learning how to elaborate the magic square using the Loubère method, it is necessary to know the parts that compose it, because although it

may not seem like it, it has a perfectly defined geometric structure that it is necessary to know and identify in order to apply the formulas that later on we will see.

Magic square structure (*odd order*).

1. Rows
2. Columns
3. Diagonals
4. Primary magic constant
5. Secondary magic constant
6. Initial position
7. Final position
8. Central position
9. Value
10. Level

Now, let´s see graphically everyone of its parts

FIGURE 2

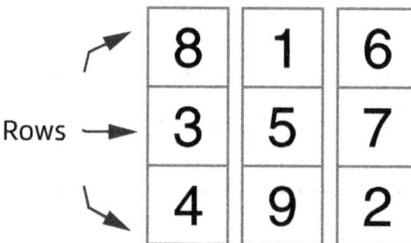

In the case of the magic square shown in this 3x3 figure, the number of rows is 3, one of 5x5 has 5 rows, one of 7x7 has 7 rows and so on, it is the horizontal distribution of the numbers.

FIGURE 3

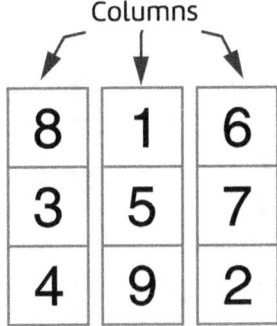

The number of columns in any magic square is always equal to the number of rows, in this case it is 3 columns.

FIGURE 4

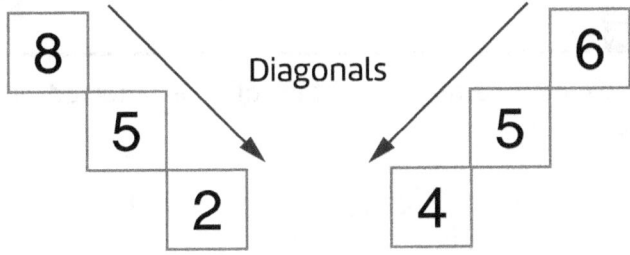

Every magic square will have only two diagonals, no matter the size.

Figure 5 illustrates the Primary Magic Constant, it is the figure obtained by adding the numbers of each row, each column and/or both diagonals.

The main objective of this book is that once you understand its structure and master the formulas that are detailed, you can obtain the magic constant that is proposed by the preferred method and apply it to the square of the size that most seems to you.

FIGURE 5

In the example shown here the Primary Magic Constant is 15

8	1	6
3	5	7
4	9	2

8+1+6=15
3+5+7=15
4+9+2=15

4+5+6=15
8+3+4=15
1+5+9=15
6+7+2=15
8+5+2=15

FIGURE 6

In this other example, a magic constant of 33 is obtained with the same magic square of 3x3

14	7	12
9	11	13
10	15	8

14+7+12=33
9+11+13=33
10+15+8=33

10+11+12=33
14+9+10=33
7+11+15=33
12+13+8=33
14+11+8=33

The secondary magic constant is a combination of 4 numbers placed symmetrically to each other with respect to the VALUE in the central position.

FIGURE 7

8	1	6
3	5	7
4	9	2

8+6+4+2+5=25

8	1	6
3	5	7
4	9	2

1+7+9+3+5=25

A magic square of order 3, or 3x3, has a maximum of 2 secondary magic constants.

In figure 8 we observe the perfect symmetry hidden by all possible secondary magic constants in a 5x5 Kumayiko

FIGURE 8

17	24	1	8	15
23	5	7	14	16
4	6	13	20	22
10	12	19	21	3
11	18	25	2	9

(1+25)+(4+22)+13=65

17	24	1	8	15
23	5	7	14	16
4	6	13	20	22
10	12	19	21	3
11	18	25	2	9

(17+9)+(15+11)+13=65

17	24	1	8	15
23	5	7	14	16
4	6	13	20	22
10	12	19	21	3
11	18	25	2	9

(8+18)+(3+23)+13=65

17	24	1	8	15
23	5	7	14	16
4	6	13	20	22
10	12	19	21	3
11	18	25	2	9

(24+2)+(10+16)+13=65

17	24	1	8	15
23	5	7	14	16
4	6	13	20	22
10	12	19	21	3
11	18	25	2	9

(5+21)+(12+14)+13=65

17	24	1	8	15
23	5	7	14	16
4	6	13	20	22
10	12	19	21	3
11	18	25	2	9

(7+19)+(6+20)+13=65

It should be noted that when talking about making a Kumayiko (magic square), it should have in mind that the magic constant to be obtained is the primary one and never the secondary one, the secondary constant is

studied here solely for the purpose of showing many of the symmetric curiosities in the geometric distribution of any Kumayiko.

The size of a magic square, or Kumayiko, we will define as Order, thus a 3x3 Kumayiko has an order of 3, a 5x5 Kumayiko has an order of 5, a 7x7 has an order of 3 and so on.

FIGURE 9

Order = *Or*

8	1	6
3	5	7
4	9	2

Order 3 Kumayiko

17	24	1	8	15
23	5	7	14	16
4	6	13	20	22
10	12	19	21	3
11	18	25	2	9

Order 5 Kumayiko

30	39	48	1	10	19	28
38	47	7	9	18	27	29
46	6	8	17	26	35	37
5	14	16	25	34	36	45
13	15	24	33	42	44	4
21	23	32	41	43	3	12
22	31	40	49	2	11	20

Order 7 Kumayiko

Every number contained in each of the cells of a Kumayiko, has established positions according to the method chosen to elaborate it, such as the Siamese Method or any of the Ascencio methods, but there are 3 most important positions that govern and determine the result of the primary magic constant, these are:

Initial position = *Pi*

Central position = *Pc*

Final position = *Pf*

Ascencio Methodology for odd Magic Squares

FIGURE 10

The figure shows how the final and initial positions are aligned in the central column and distanced from the central position in a symmetrical and equidistant manner, the central position plays a role key and is fundamental piece in obtaining the results of the primary magic constant.

Initial position

30	39	48	1	10	19	28
38	47	7	9	18	27	29
46	6	8	17	26	35	37
5	14	16	25	34	36	45
13	15	24	33	42	44	4
21	23	32	41	43	3	12
22	31	40	49	2	11	20

Central position — points to 25 (center). Final position — points to 49.

VALUE: It is the digit or number that occupies a cell at a given moment, while the position is absolute, the value is relative and changes indefinitely with each new and different Kumayiko.

FIGURE 11

In these two examples we see how the same 5x5 Kumayiko contains different values to obtain different magic constants, but the numbers of the positions (small numbers) remain fixed.

17	24	1	8	15
20	27	4	11	18
23	5	7	14	16
26	8	10	17	19
4	6	13	20	22
7	9	16	23	25
10	12	19	21	3
13	15	22	24	6
11	18	25	2	9
14	21	28	5	12

17	24	1	8	15
11	18	-5	2	9
23	5	7	14	16
17	-1	1	8	10
4	6	13	20	22
-2	0	7	14	16
10	12	19	21	3
4	6	13	15	-3
11	18	25	2	9
5	12	19	-4	3

Finally we will see the definition of **Level** in the following figure

FIGURE 12

All Kumayikos are made up of a series of cells grouped concentrically as if they were rings around the central cell in multiples of 8, level 1 consists of 8 cells, level 2 consists of 16 cells, level 3, 24 cells and so on.

A Kumayiko of order 3, consists of 1 level, a Kumayiko of order 5, consists of 2 levels, one of order 7, as shown in the figure below, consists of 3 levels, and a Kumayiko of order 9, consists of 4 levels.

NOTE: The central position it is not considered a a level by itself.

	30	39	48	1	10	19	28	
Level 2	38	47	7	9	18	27	29	Level 3
	46	6	8	17	26	35	37	
	5	14	16	25	34	36	45	
Level 1	13	15	24	33	42	44	4	
	21	23	32	41	43	3	12	
	22	31	40	49	2	11	20	

2. Mathematical curiosities

One feature that caught my attention when I began to study magic squares, or Kumayikos, as we will call them throughout this book, is that they present a series of amazing geometric arrangements that are surprising for their perfect symmetry.

One of them is the mathematical relationship that exists between the primary magic constant and the secondary magic constant, which is defined by a proportion factor determined by the size of the square between the number 5, let's see it graphically to understand it better.

(Secondary magic constant) x (Or/5) = Primary magic constant

FIGURE 13

Size of the square = Or = 7

30	39	48	1	10	19	28
38	47	7	9	18	27	29
46	6	8	17	26	35	37
5	14	16	25	34	36	45
13	15	24	33	42	44	4
21	23	32	41	43	3	12
22	31	40	49	2	11	20

Primary magic constant
22+23+24+25+26+27+28=175

30	39	48	1	10	19	28
38	47	7	9	18	27	29
46	6	8	17	26	35	37
5	14	16	25	34	36	45
13	15	24	33	42	44	4
21	23	32	41	43	3	12
22	31	40	49	2	11	20

Secondary magic constant
30+28+22+20+25=125

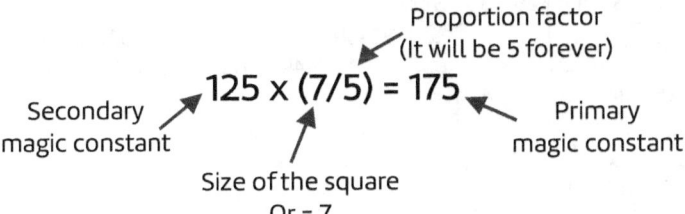

Proportion factor
(It will be 5 forever)

$$125 \times (7/5) = 175$$

Secondary magic constant — Primary magic constant

Size of the square
Or = 7

FIGURE 14

Another strange and amazing curiosity is the relationship that exists between the number of cells contained in each level with the value of the central number, in the case of a 7x7 Kumayiko, which consists of 3 levels, we can appreciate this mathematical relationship in figure 14.

30	39	48	1	10	19	28
38	47	7	9	18	27	29
46	6	8	17	26	35	37
5	14	16	25	34	36	45
13	15	24	33	42	44	4
21	23	32	41	43	3	12
22	31	40	49	2	11	20

Level 1: 8 cells

25 x 8 = 200

8+17+26+16+34+24+33+42 = 200

30	39	48	1	10	19	28
38	47	7	9	18	27	29
46	6	8	17	26	35	37
5	14	16	25	34	36	45
13	15	24	33	42	44	4
21	23	32	41	43	3	12
22	31	40	49	2	11	20

Level 2: 16 cells

25 x 16 = 400

47+7+9+18+27+35+36+44+
3+43+41+32+23+15+14+6 = 400

30	39	48	1	10	19	28
38	47	7	9	18	27	29
46	6	8	17	26	35	37
5	14	16	25	34	36	45
13	15	24	33	42	44	4
21	23	32	41	43	3	12
22	31	40	49	2	11	20

Level 3: 24 cells

25 x 24 = 600

30+39+48+1+10+19+28+29+37+45+4+12+
20+11+2+49+40+31+22+21+13+5+46+38 = 600

Ascencio Methodology for odd Magic Squares

Incredible and regardless of the size of the Kumayiko in question, each level adds 8 cells to the lower level and the sum of that level is the multiple of the value of the central position by the number of cells contained in each level with a perfect and symmetrical progression numerical.

Perhaps the most amazing feature of the Kumayikos is the perfect symmetry and balance that is maintained throughout its structure with respect to the central number, that is, if we take any pair of numbers that are located at 180 degrees with respect to each other, it will always give as a result twice the value of the central number.

FIGURE 15

30	39	48	1	10	19	28
38	47	7	9	18	27	29
46	6	8	17	26	35	37
5	14	16	25	34	36	45
13	15	24	33	42	44	4
21	23	32	41	43	3	12
22	31	40	49	2	11	20

20+30 = 50

30	39	48	1	10	19	28
38	47	7	9	18	27	29
46	6	8	17	26	35	37
5	14	16	25	34	36	45
13	15	24	33	42	44	4
21	23	32	41	43	3	12
22	31	40	49	2	11	20

22+28 = 50

30	39	48	1	10	19	28
38	47	7	9	18	27	29
46	6	8	17	26	35	37
5	14	16	25	34	36	45
13	15	24	33	42	44	4
21	23	32	41	43	3	12
22	31	40	49	2	11	20

5+45 = 50

30	39	48	1	10	19	28
38	47	7	9	18	27	29
46	6	8	17	26	35	37
5	14	16	25	34	36	45
13	15	24	33	42	44	4
21	23	32	41	43	3	12
22	31	40	49	2	11	20

27+23 = 50

30	39	48	1	10	19	28
38	47	7	9	18	27	29
46	6	8	17	26	35	37
5	14	16	25	34	36	45
13	15	24	33	42	44	4
21	23	32	41	43	3	12
22	31	40	49	2	11	20

47+3 = 50

30	39	48	1	10	19	28
38	47	7	9	18	27	29
46	6	8	17	26	35	37
5	14	16	25	34	36	45
13	15	24	33	42	44	4
21	23	32	41	43	3	12
22	31	40	49	2	11	20

15+35 = 50

30	39	48	1	10	19	28
38	47	7	9	18	27	29
46	6	8	17	26	35	37
5	14	16	25	34	36	45
13	15	24	33	42	44	4
21	23	32	41	43	3	12
22	31	40	49	2	11	20

31+19 = 50

30	39	48	1	10	19	28
38	47	7	9	18	27	29
46	6	8	17	26	35	37
5	14	16	25	34	36	45
13	15	24	33	42	44	4
21	23	32	41	43	3	12
22	31	40	49	2	11	20

39+11 = 50

30	39	48	1	10	19	28
38	47	7	9	18	27	29
46	6	8	17	26	35	37
5	14	16	25	34	36	45
13	15	24	33	42	44	4
21	23	32	41	43	3	12
22	31	40	49	2	11	20

6+44 = 50

An amazing mathematical anecdote...

Carl Friedrich Gauss was a prominent German astronomer, physicist, and mathematician born in 1777 who is credited with greatly advancing number theory, mathematical analysis, differential geometry, statistics, and algebra.

As a child, he showed some very noticeable signs of his genius in mathematics, such as when he corrected his father's ledger before he was 10 years old.

Another great anecdote was when at the age of nine, in their arithmetic class, the teacher assigned them the task of adding the numbers from 1 to 100, with the sole purpose of keeping them busy for a long time, which did not happen with Carl, that he managed to have the answer in a very short time, because he realized that making pairs of numbers with the first and the last, the second and the penultimate, the third and the penultimate and so on, all the pairs resulted in the 101, that is, 100+1 = 99+2 = 98+3 = 97+4, = 96+5 = 95 + 6..... 51+50 = 101, so in 100 numbers we had 50 pairs of 101, so the only thing he had to do was multiply those figures, resulting in 101x50 = 5050, without wanting to, he had discovered the formula to obtain the sum of the terms of an arithmetic progression, when he was just nine years old !!

FIGURE 16

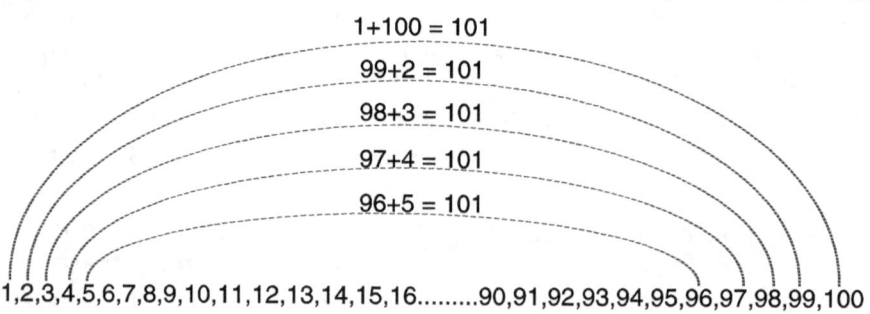

Imaginary distribution of pairs of numbers as visualized by the prodigy Cchild Carl Friedrich Gauss at the age of 9 to obtain the sum of the numbers from 1 to 100.

Incredibly, somewhat similar to Gauss's formula for arithmetic progression, the value of the central number in a Kumayiko plays a crucial role thet allows us to obtain the sum of the arithmetic progression of all the numbers contained in the magic square when multiplied by the total number of cells contained in a Kumayiko.

In figure 17 we observe 2 examples with Kumayikos of different sizes to obtain the sum of all their numbers by simply multiplying their central number by the total of their cells.

FIGURE 17

17	24	1	8	15
23	5	7	14	16
4	6	**13**	20	22
10	12	19	21	3
11	18	25	2	9

13 x 25 = 325

30	39	48	1	10	19	28
38	47	7	9	18	27	29
46	6	8	17	26	35	37
5	14	16	**25**	34	36	45
13	15	24	33	42	44	4
21	23	32	41	43	3	12
22	31	40	49	2	11	20

25 x 49 = 1225

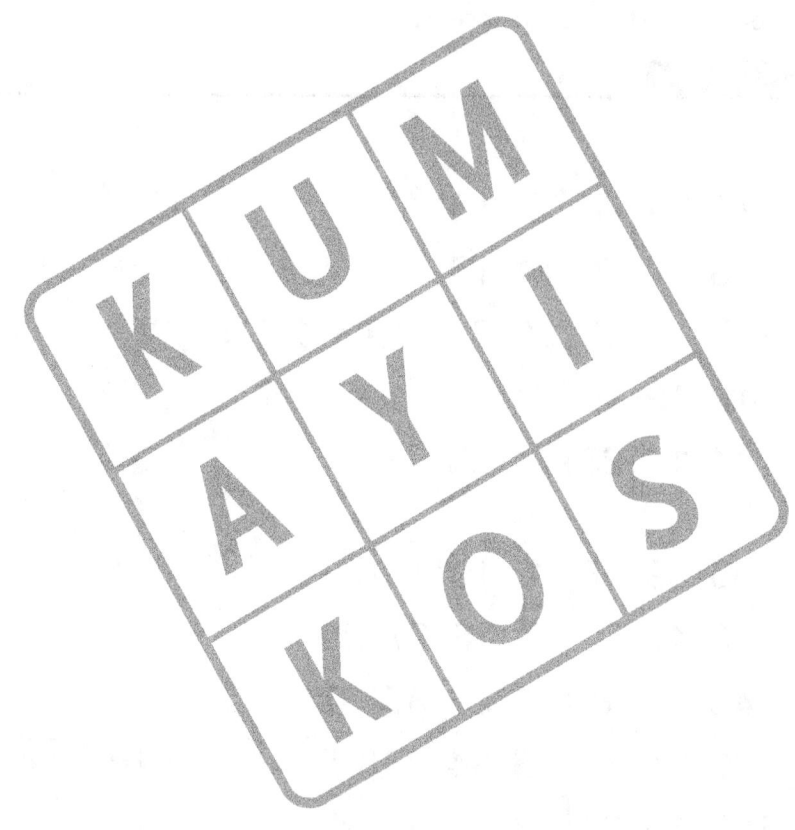

Second Part...

¡ Let's do it !

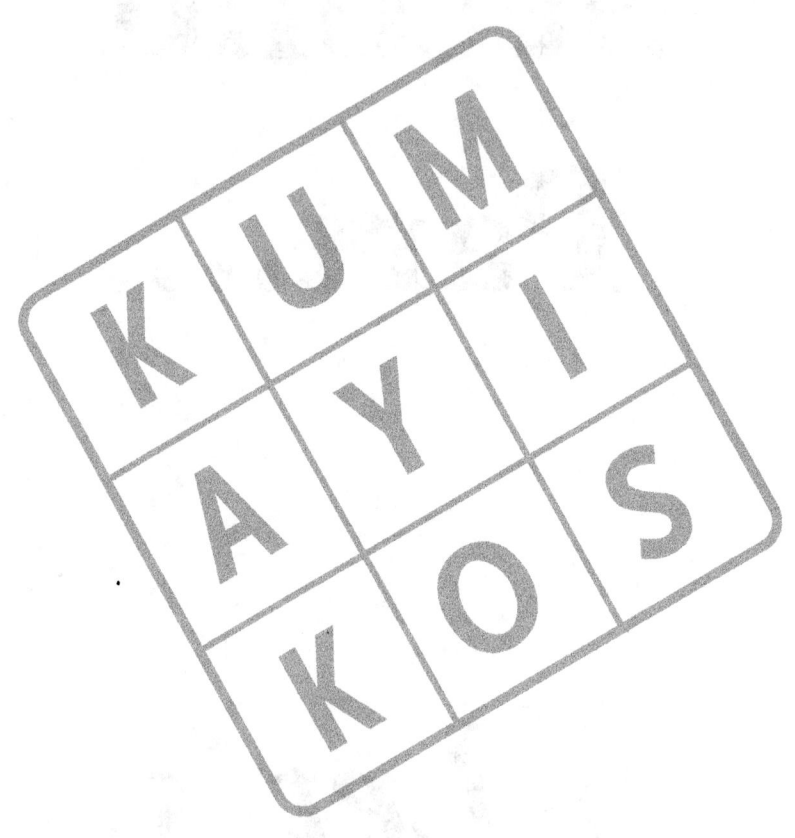

3. De la Loubère Method

Siamese Method

Having studied a bit of the basics about the structure of an odd order magic square, we can proceed to study the method that my father taught me on a napkin over 30 years ago, named after the French diplomat Simon de la Loubère who published it a few years before 1700, back in 1691, shortly after returning from the Kingdom of Siam (now Thailand), which is why it is also known as the Siamese Method.

This method, as himself relates, was taught to him on one of the ships that took him back to Europe, a doctor who had learned it in India, he documented it for two reasons, one, for having fascinated him, and another, because it was the orders of King Louis XIV, to write and document everything he saw on his journey, and one of the unique and fascinating things he saw on his travels was precisely the method for making magic squares of odd order, in fact the square that he published in volume II of the "Kingdom of Siam" was one of 5x5.

Personally, I don't know how my father learned it, or who taught it to him, but somehow I know that it was empirically transmitted to him by someone he knew through an endless chain that began many centuries ago in the East. medium and despite lacking any technical utility in real life, has fascinated many people including great mathematicians in history such as Pascal, Benjamin Franklin, Fermat, Leibnitz, Euler and others until reaching this humble writer who writes the pages that you have in your hands, who is far from being a mathematician, but that doesn't stop him from writing this book.

To fill in our first Kumayiko, we will use a 5x5 square matrix, positioning our first number in the Initial position cell and starting from there using some simple rules that we will see step by step throughout the following pages.

Rule no. 1
The first number is placed in the starting position cell that is in the center of the top row.

FIGURE 18

In this case, as an initial exercise, we will begin by placing the number 1 in the initial position, but we could have started with any number.

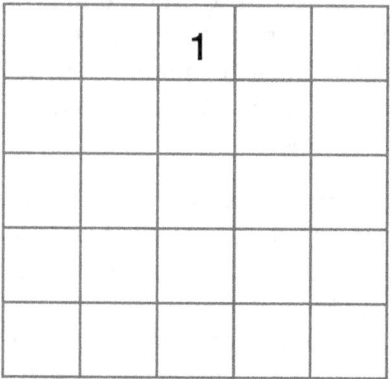

Rule no. 2
Any consecutive number of the number placed in the upper line, except for the right corner, will be placed in the cell of the lower line one column to the right

FIGURE 19

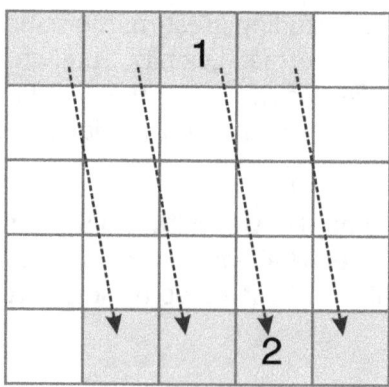

Rule no. 3

Whenever the cell that is in the upper line to the right is free, the consecutive of any number will be placed there, in this case the cell that is above and to the right of the 2 was free, the next number will be placed there, in this case the no. 3

FIGURE 20

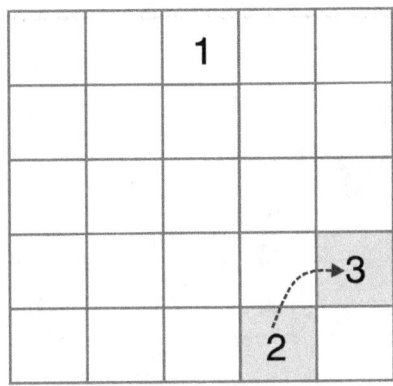

Rule no. 4

All consecutive numbers placed in the far right column, with the exception of the cell in the upper right corner, will be placed one line up in the far left column.

FIGURE 21

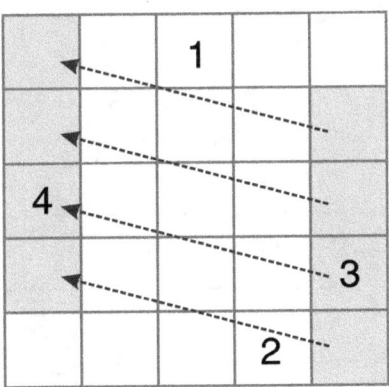

Rule no. 5

Number 5 was placed above and to the right of number 4 following rule no. 3, but to place the number 6, you cannot follow rule no. 3 since it is occupied by the number 1, in this case rule no. 5, which means that when the number on the top right is occupied, then the number will be placed in the cell immediately below..

FIGURE 22

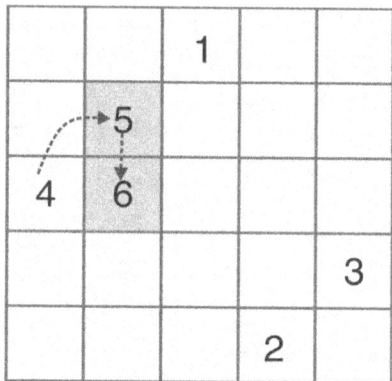

Numbers 7 and 8 can be placed following rule no. 3 , since the cells above and to the right are empty, and the number 9 is placed following rule no. two.

FIGURE 23

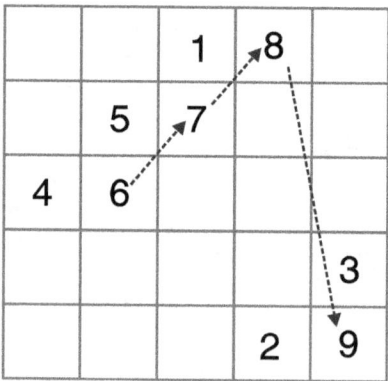

Rule no. 6

To explain rule no. 6, we will place first the numbers 10, 11, 12, 13, 14 and 15.

The number 10 is placed following rule no. 4, the number 11 is placed following rule no. 5 and the numbers 12, 13, 14 and 15 are placed following rule no. 3.

Now we will see that the number 16 is placed just below the number 15, that is, rule no. 6 specifies that every number following the number placed in the upper right corner is always placed just below that cell..

FIGURA 24

		1	8	15
	5	7	14	16
4	6	13		
10	12			3
11			2	9

These simple and easy 6 rules summarize the method that Simon de la Loubère learned on his trip to the Kingdom of Siam and he published it at the end of 1600, with these 6 rules you will be able to make odd magic squares as big as you want, just practicing a little bit.

Applying the rules described, we can proceed to finish placing the remaining numbers to finish our first Kumayiko.

The 17 is placed according to rule no. 4, the 18 is placed with rule no. 2, 19 and 20 are placed following rule no. 3 which is up and to the right, 21 is placed following rule no. 5, 22 is placed following rule no. 3, 23 is placed according to rule no. 4, the 24 is placed following rule no. 3 and finally 25 is placed according to rule number two.

In the figure we see a completely finished Kumayiko, with a summary of the rules that were applied for each of its values.

FIGURE 25

Rule 1:
-> 1
Rule 2:
-> 18, 25, 2 y 9
Rule 3:
-> 3, 5, 7, 8, 12, 13, 14, 15, 19, 20, 22 y 24
Rule 4:
-> 10, 4, 23 y 17
Rule 5:
-> 6, 11 y 21
Rule 6:
-> 16

17	24	1	8	15
23	5	7	14	16
4	6	13	20	22
10	12	19	21	3
11	18	25	2	9

4. Central Position Method

The constant magic that you choose

It is with this method that the elaboration of Kumayikos becomes more interesting, just by introducing the value of the central number to the equation, you will be able to obtain total control of the magic constant, since we will be able to choose it at will and elaborate the Kumayiko that you want.

Suppose you want to make a 5x5 Kumayiko whose magic constant is 240, that is, when adding the values of the columns, diagonals and rows, we get 240 as a result, we simply proceed to divide this value by the size of the Kumayiko, which means dividing the 240 by 5 to obtain a value of 48 that will be placed in the central position, then we calculate the distance that this number is from the initial position and proceed to complete the Kumayiko with the de la Loubere method.

Let´s see it step by step at detail:

Step 1:
The magic constant that we chose, in this case 240, is divided by the size of the square and the result is placed in the central position, in this case, 48.

FIGURE 26

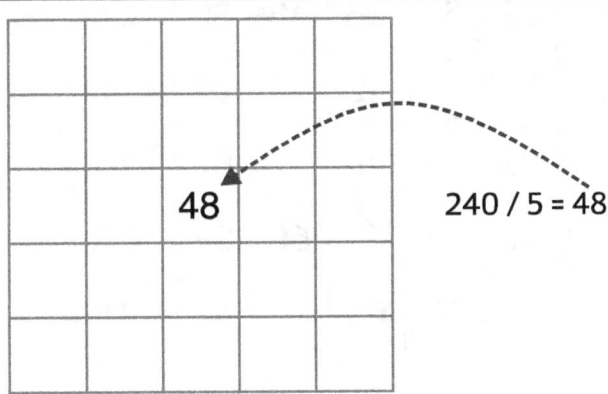

Step 2:
We subtract number 1 from the total number of cells in the Kumayiko and

divide it by 2 to know the value of the number from the initial position.

FIGURE 27

In the case of Kumayiko of order 5, the total number of cells is 25

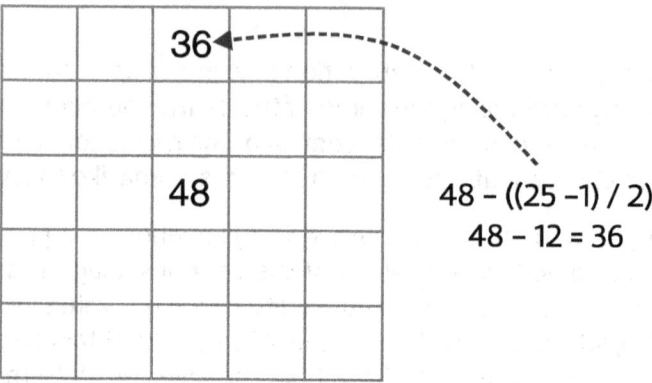

48 – ((25 –1) / 2)
48 – 12 = 36

Step 3:
Once we got the value of the starting position number, we can proceed to complete the Kumayiko using the de la Loubère method to check that the value of the magic constant is 240.

FIGURE 28

We can validate and see that the magic constant is 240 !

52	59	36	43	50	= 240
58	40	42	49	51	= 240
39	41	48	55	57	= 240
45	47	54	56	38	= 240
46	53	60	37	44	= 240

Let's do other exercise, but now with a 3x3 Kumayiko to get the same magic constant of 240.

$$240 / 3 = 80$$

$$80 - ((9-1) / 2)$$

$$80 - 4 = 76$$

FIGURE 29

Amazing, is not it? Same magic constant of 240 with different sized Kumayikos.

83	76	81	= 240
78	80	82	= 240
79	84	77	= 240

Just remember these simple steps:

The magic constant is divided by the size of the Kumayiko and that value is placed in the central position, then the total number of cells is subtracted by 1 and divided by 2, the result is subtracted from the value of the central position and that number is placed in the initial position, and that's it, the next thing is to complete the Kumayiko with the Loubère method that we learned in chapter 3.

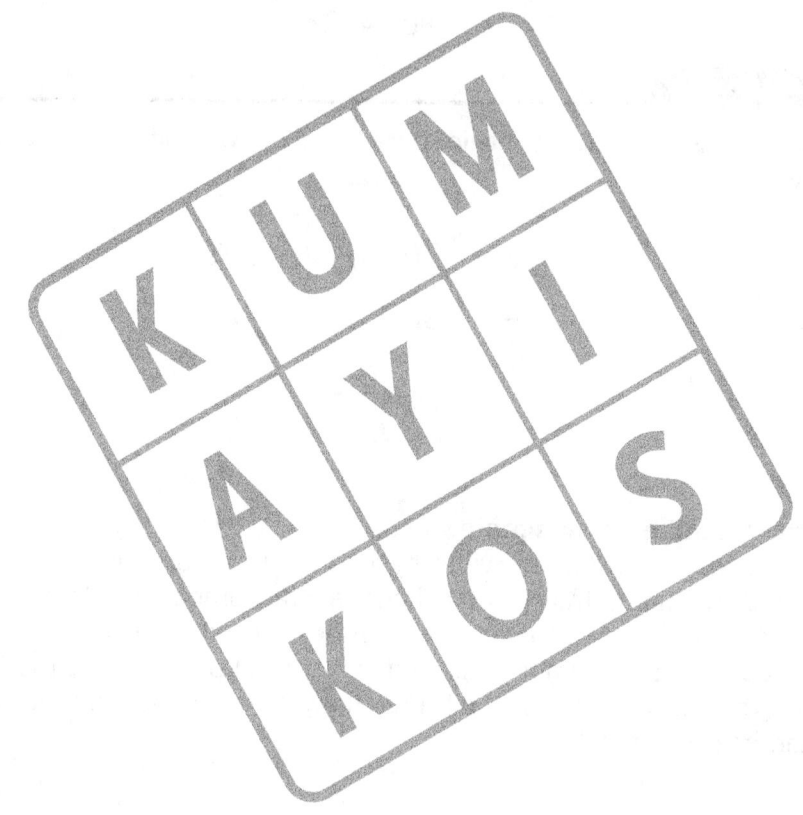

5. 180 degree pairwise method

If Carl Friedrich Gauss lived...

De La Loubère's method offers a method to elaborate magic squares by placing all the numbers of the series sequentially, but with the method we are going to study now, the placement of the numbers will be in pairs in some similar way as Carl Friedrich visualized them in elementary school when the teacher asked them to obtain the sum of all the numbers from 1 to 100, the first with the last, the second with the penultimate, the third with the antepenultimate and so on until the pairs are completed.

Thus, for example, for a 5x5 Kumayiko, in which we have 25 numbers, the numbers will be grouped in the following pairs:

(1-25) (2-24) (3-23) (4-22) (5-21) (6-20) (7-19) (8-18) (9-17) (10-16) (11-15) y (12-14), *the last number, 13, is not paired with any other, since it is the number that occupies the central position (Pc)*

With this grouping we can understand where the name of the "Pairwise" method comes from, but... what does the term 180° refer to?

I named it 180 degree pairwise method because all the pairs that we have grouped will be arranged at 180 degrees with respect to each other, taking the central position of the Kumayiko as the geometric center.

As you do it, you will realize that you must master the traditional filling of the Kumayiko using the La Loubère method, since to place a number of each pair, you must rely on such method to know where the next number will be placed.

Without preamble, let's proceed to make a 5x5 Kumayiko whose initial number is 1 and the increments are likewise 1, in figure 29 we can see it.

The rule is simple, the first number of each pair will be placed following the Siamese or La Loubère method, as you like to name it, and the second number of the pair will be placed at 180 degrees, one after another until finished.

FIGURE 30

① The first number of the pair is placed in the starting position as we saw it in the Siamese method

First Pair

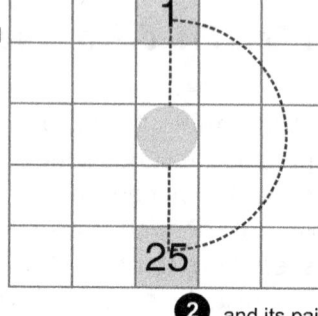

② and its pair, the 25, to 180 degree

④ ...and it´s pair, number 24, is placed 180 degrees

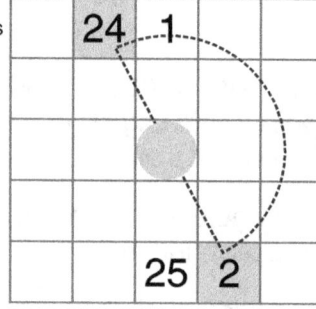

Second Pair

③ number 2 is placed following the Siamese method...

⑥ and its pair, number 23 placed at 180 degrees

Third Pair

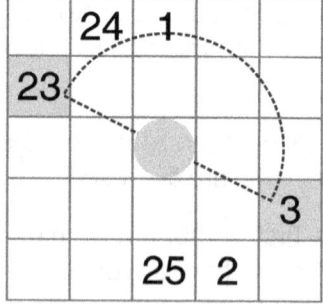

⑤ number 3 is placed following the Siamese method...

Figura 30a

Fourth pair

⑦ the number 4 is placed following the siamese method

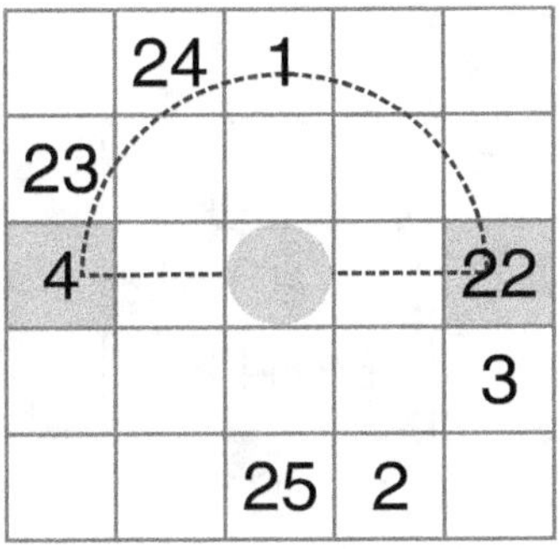

⑧ and its pair, the number 22, 180 degrees from 4

	24	1		
	23			
4			22	
			3	
	25	2		

⑨ the 5, is placed up and to the right of 4, following the Siamese method

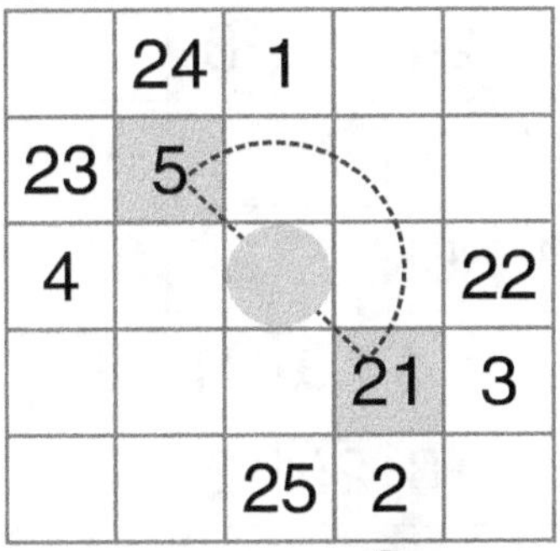

	24	1		
	23	5		
4			22	
		21	3	
	25	2		

Fifth pair

⑩ its pair, 21, is placed at 180 degrees of the number 5

Sixth Pair

⑪ the 6 is placed below 5, following the Siamese method

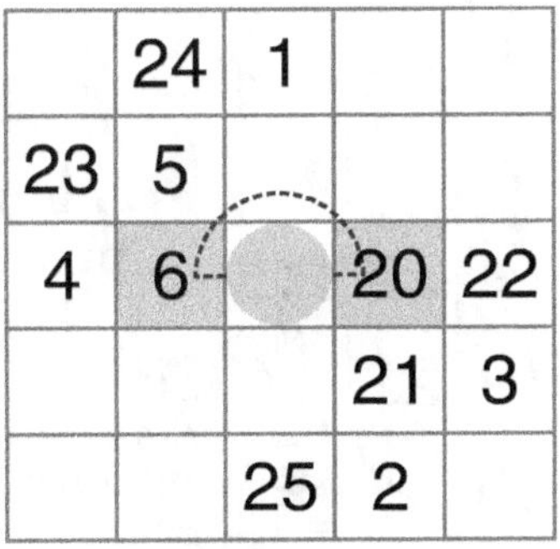

	24	1		
	23	5		
4	6		20	22
		21	3	
	25	2		

⑫ and its pair, the 20, just at 180 degrees from 6

FIGURA 30b

	24	1		
23	5	7		
4	6		20	22
		19	21	3
		25	2	

(13) the number 7 is placed above and to the right of 6

Seventh pair

(14) and its pair, the number 19, 180 degrees from 7

	24	1	8	
23	5	7		
4	6		20	22
		19	21	3
	18	25	2	

(15) the 8, is placed above and to the right of 7, following the Siamese method

Eigth Pair

(16) and its pair, 18 is placed 180 degrees from 8

(18) and its pair, 17, just at 180 degrees in the opposite corner

17	24	1	8	
23	5	7		
4	6		20	22
		19	21	3
	18	25	2	9

Nineth pair

(17) the 9 is placed in the lower right corner

Ascencio Methodology for odd Magic Squares

FIGURA 30c

Tenth Pair

17	24	1	8	
23	5	7		16
4	6		20	22
10		19	21	3
	18	25	2	9

(19) the 10, is placed following the rule #4 of siamese method

(20) and its pair, number 16, at 180 degrees from 10

Eleventh pair

17	24	1	8	15
23	5	7		16
4	6		20	22
10		19	21	3
11	18	25	2	9

(21) the 11 is placed below 10 according to Siamese method

(22) 15 is placed at 180 degrees from 11

Twelfth pair

17	24	1	8	15
23	5	7	14	16
4	6	13	20	22
10	12	19	21	3
11	18	25	2	9

(23) the 12 is placed up and to the right from 11 following rule #3 of the siamese method

(24) the last pair is place at 180 degrees from 12,

(25) and finally the central number is placed in the geometric center position of the Kumayiko.

33

Well, we have completed our first Kumayiko based on the 180 degree pair wise method, which gives the same result as the La Loubère method, with the difference that using the pair method we can appreciate the amazing property of perfect symmetry that is maintains the length and breadth of the Kumayiko at the same time that it is being made, regardless of the size in question, it is surprising to observe how each pair that is aligned is exactly twice the value that the central number occupies.

6. Method for solving the Lu-Shu theorem

Infinite combinations

So far we have learned to make Kumayikos through various methods, but all having one thing in common, that they start from the initial position (Pi), but what if we randomly took two numbers as if they were dices and placed them in the same way way in random positions within the Kumayiko? Would it be possible to generate a Kumayiko starting from a condition like the one described?

There is a theorem that literally goes like this:

"It is possible to generate infinite magic squares from two given numbers randomly placed in any of the cells of a magic square"

But one thing is a theorem and another thing is its validation, so without preamble, let's go not only to the proof but also to a valid and systematized method for obtaining a Kumayiko starting from this mathematical statement

We are going to choose two random numbers, say 62 and 77, as well as two random positions, say absolute position 5 and absolute position 8 in a 3x3 Kumayiko

FIGURE 31

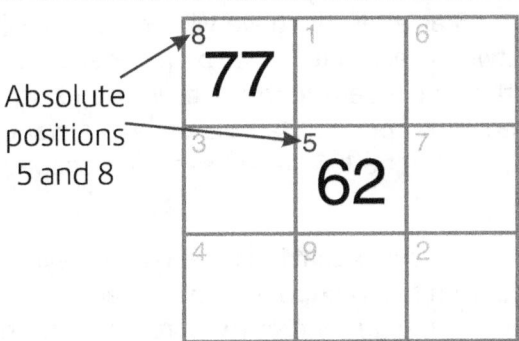

To solve this random combination, we must ask and answer 3 simple questions:

1.- What are the absolute positions in which the numbers are placed? Answer: 5 and 8

2.- What difference do we get from subtracting the values of the absolute positions? Answer: 8 − 5 = 3

3.- What difference do we get from subtracting the 2 numbers placed in the absolute positions? Answer: 77 − 62 = 15
Now let's see it graphically in the following figure

FIGURE 32

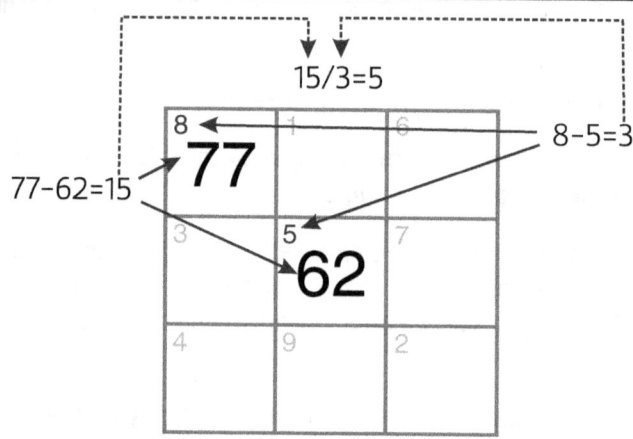

In a few simple words, this means that there are 3 cells of difference between the absolute position of one number and another and it is precisely those 3 cells in which we will divide 15, which is the difference between 77 and 62, getting the 5 as a result, so we will add 5 to 62 (the number 5 is the increment between each cells) and so on to each cell to complete the Kumayiko using the La Loubère method, as in the example we place the number 62 in the central position, as we know in advance that the value of the magic constant will be 186, as we learned in chapter 4 on the center position method.

Once completed, we check that it is indeed a Kumayiko, a perfect magic square, generated from two random numbers placed in random positions, so we can say that the Lu-Shu Theorem was true, the only thing missing was a method valid and easy to check like the one you just learned.

FIGURE 33

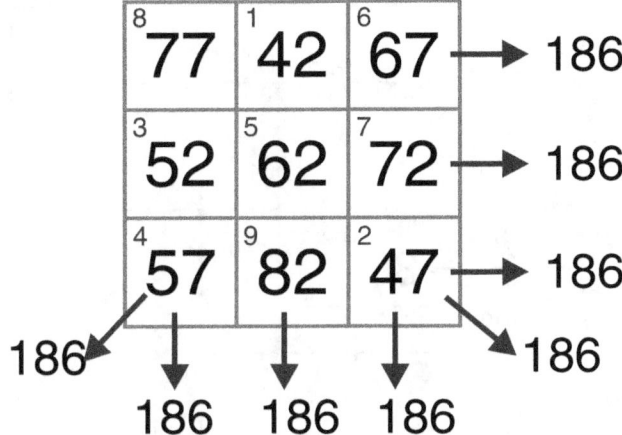

We have verified that the Lu-Shu Theorem is valid and learned to solve it using a 3x3 Kumayiko, now we are going to generate a Kumayiko using a 5x5 matrix and for this we will use the numbers 12 and 36 in positions 3 and 6.

FIGURE 34

17	24	1	8	15
23	5	7	14	16
4	⁶36	13	20	22
10	12	19	21	³12
11	18	25	2	9

Let's proceed to carry out the necessary steps to be able to obtain the necessary values and place them in the empty cells to complete the Kumayiko through the Lu-Shu Theorem

FIGURE 35

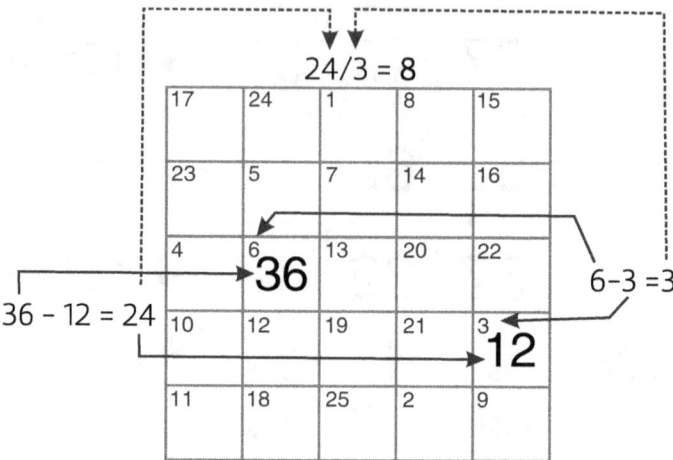

The number obtained in this case was 8, and the number of steps required to reach the number 36 is 3, these numbers will be placed using the de la Loubère method.

FIGURE 36

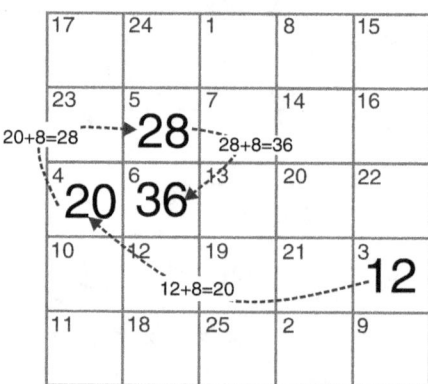

Once the Kumayiko is completed, we can validate that it is indeed a perfect magic square, since the magic constant is 460 in the sum of its diagonals, columns and rows, this same constant can be obtained using the central position method, and it is simply multiplying the number of the central position that in this case gave us 92 by the number of the Kumayiko, which is 5 and we obtain the magic constant of 460, as wen can see in figure 36.

FIGURE 37

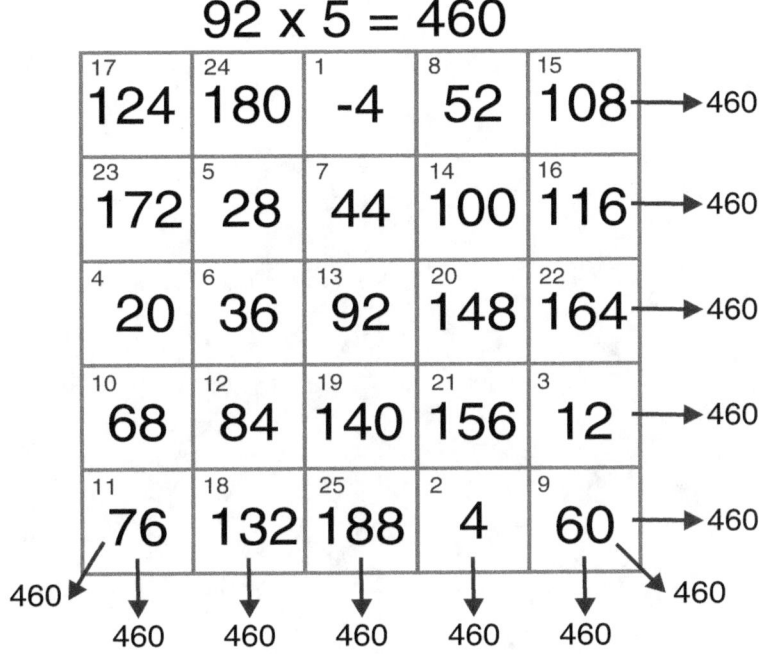

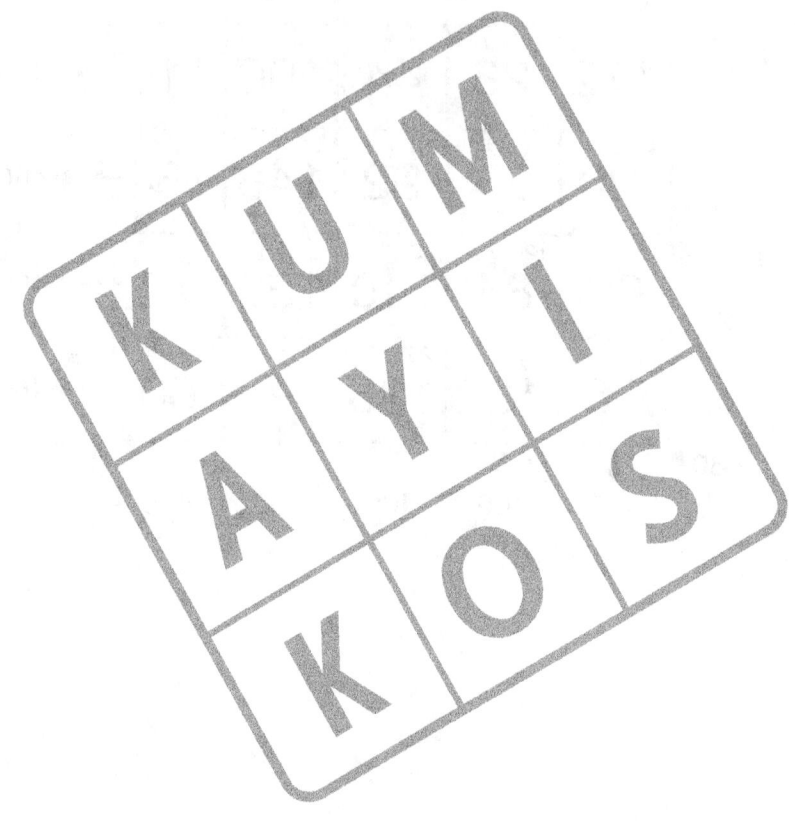

7. Multiplication Method

The cherry of the cake!

I intentionally leave this chapter for the end, because for its elaboration, it is essential to master the La Loubère method, the method that has dominated the world of magic squares for hundreds of years..

This new multiplicative method is a watershed in the history of magic squares, it breaks the pattern that for more than 2,200 years had surrounded squares obtaining their magic constant only by addition, this method with which I crown my work and that I put into your hands, it is the first in history to allow you in a simple way, based on logarithmic laws, to obtain a magic constant through MULTIPLICATIONS and not only by additions, mathematically you will be able to express it through exponents and radical indices, but for its solution, I recommend the use of a scientific calculator, since the numbers you will get will consist of several decimal numbers and figures, very, very long.

The formula that determines these values is basically the following:

FIGURE 38

$$\sqrt[Cm]{X^{Pos}}$$

For the elaboration of a Kumayiko we could say that 2 Kumayikos are going to be integrated into one, it seems complicated, but believe me that it is surprisingly simple and in some way similar to the approach of the variables in the Lu-Shu Theorem, this multiplicative method consists of 3 steps:

1.- The number that we want to obtain through multiplications is placed in each cell, for this example we will use the number 88.

FIGURE 39

88	88	88
88	88	88
88	88	88

Then we proceed to elaborate a conventional Kumayiko with a magic constant of 15 and we place in the upper right part of each radicand as an exponent, that is, we are going to raise 88 to the power of each number placed in the corresponding cell.

FIGURE 40

88^8	88^1	88^6
88^3	88^5	88^7
88^4	88^9	88^2

Finally, the magic constant resulting from the Kumayiko by sums that we place in figure 39 is placed in each cell, but we will use the value as an index of the radical, and that is all, when multiplying the values obtained from each cell by columns, rows and diagonals we will obtain the magic constant 88 through multiplications.

FIGURE 41

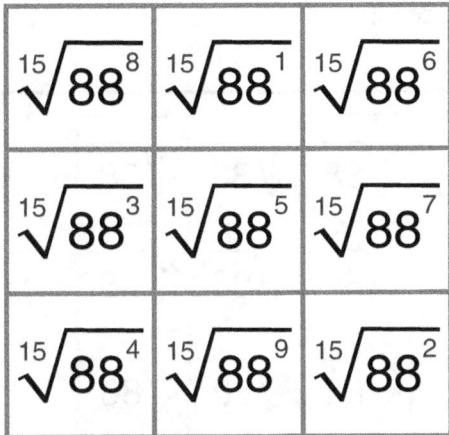

It is clear that to solve this type of Kumayiko, the use of a scientific calculator is required. In the following figure we can observe the values obtained in each cell.

FIGURE 42

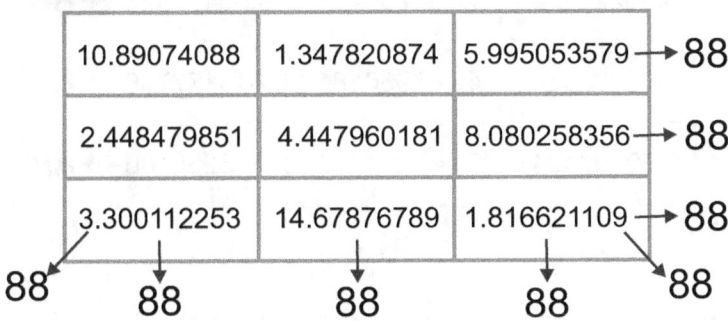

It's amazing, isn't it? We have elaborated the first Kumayiko obtaining the magic constant by MULTIPLICATIONS and not by additions. We have broken the pattern that dominated the world of magic squares for more than 2,200 years !

If you pay attention to the Kumayiko by multiplications that we have just elaborated, you will notice that we used to elaborate a Kumayiko by additions whose magic constant was 15, but we are not limited to that value, you will be able to occupy conventional Kumayikos with the magic constant that you choose, let's try by example get the same number 88 by multiplications using a magic constant of 21.

FIGURE 43

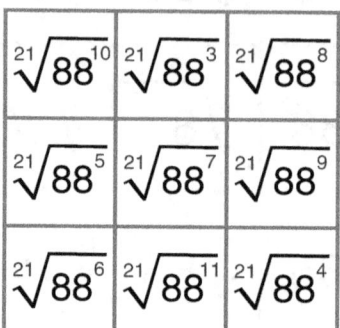

Breaking down the values of each cell, we obtain the results shown in figure 43.

FIGURE 44

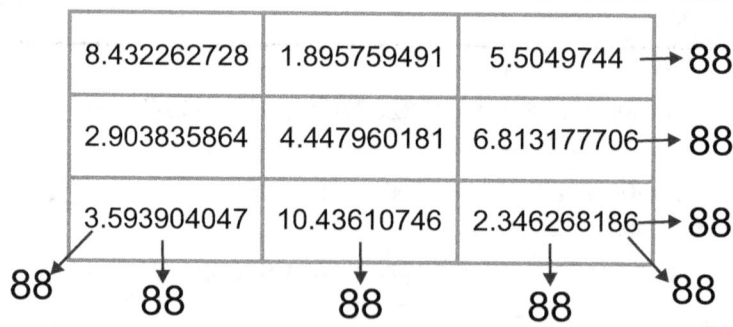

Plain and simply amazing, isn't it? Just by varying the Kumayiko by sums that acts as an exponent, as well as its magic constant, we can generate infinite combinations of the same magic constant by multiplications using a Kumayiko of the same size..

In figure 44 we can graphically appreciate the 3 parts that make up a multiplicative Kumayiko.

FIGURE 45

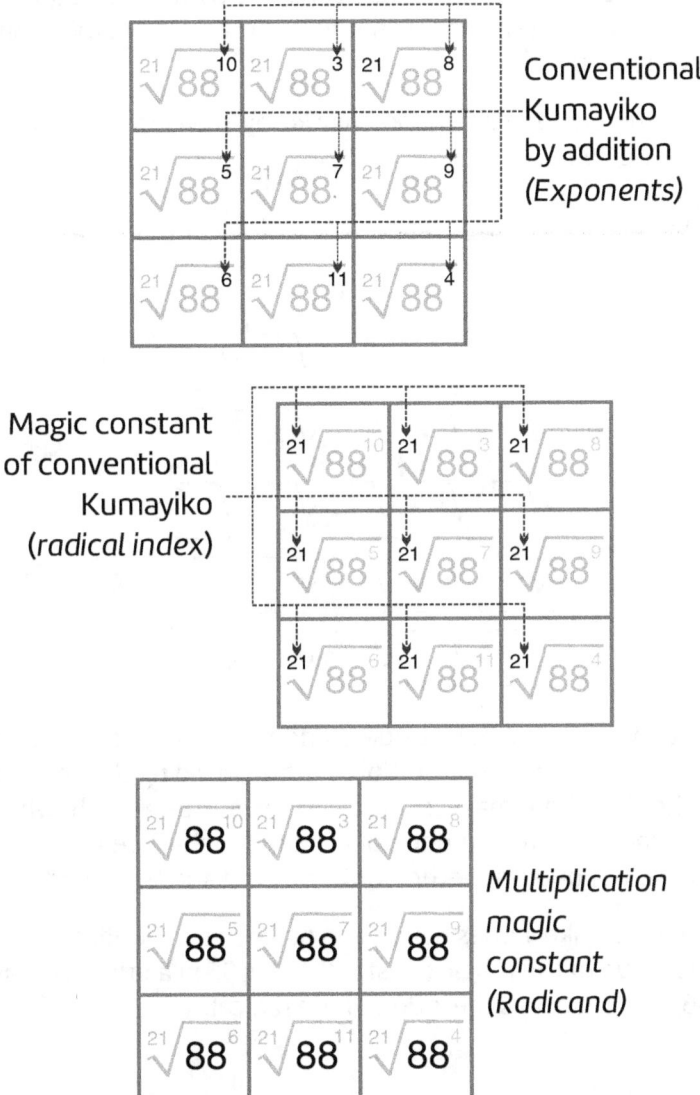

Conventional Kumayiko by addition (Exponents)

Magic constant of conventional Kumayiko (radical index)

Multiplication magic constant (Radicand)

Before moving on to the next chapter, how about we apply what we have learned so far and see how effective this method is on a 5x5 Kumayiko and see if the same rules apply not only to the primary magic constant, but also to the secondary magic constant. that we studied in the first pages of this book, to be exact on page 8, figures 7 and 8.

To make the 5x5 Kumayiko, we will use a Kumayiko by addition with a magic constant of 40 and the magic constant by multiplications that we want to obtain is 360 (remember that you can use the values you like for these two variables).

Let's start by elaborating the Kumayiko by sums that will serve as an exponent of the 360.

FIGURE 46

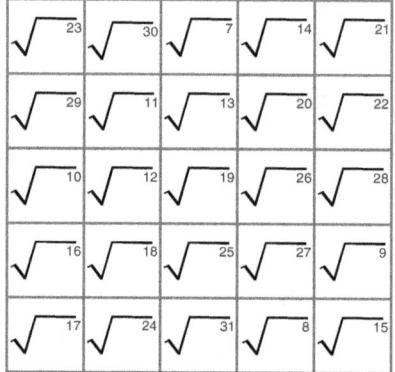

At this point we can already understand that to obtain the magic constant of 95 we use the central position method, dividing the constant 95 by 5, which is the size of the magic square, giving us 19 as a result, which occupies the central position, and once placing that value, we proceed to fill it through the de La Loubère method both forward and backward.

Once the Kumayiko values have been placed as exponents, we proceed to place the value of the magic constant that is 95 in all the cells, as shown in figure 46, this will act as the index of the radical.

FIGURE 47

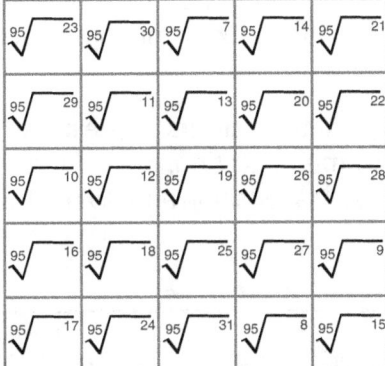

And finally we place the magic constant that we want to obtain in each of the cells, as a radicand of the radical.

FIGURE 48

$\sqrt[95]{360}^{23}$	$\sqrt[95]{360}^{30}$	$\sqrt[95]{360}^{7}$	$\sqrt[95]{360}^{14}$	$\sqrt[95]{360}^{21}$
$\sqrt[95]{360}^{29}$	$\sqrt[95]{360}^{11}$	$\sqrt[95]{360}^{13}$	$\sqrt[95]{360}^{20}$	$\sqrt[95]{360}^{22}$
$\sqrt[95]{360}^{10}$	$\sqrt[95]{360}^{12}$	$\sqrt[95]{360}^{19}$	$\sqrt[95]{360}^{26}$	$\sqrt[95]{360}^{28}$
$\sqrt[95]{360}^{16}$	$\sqrt[95]{360}^{18}$	$\sqrt[95]{360}^{25}$	$\sqrt[95]{360}^{27}$	$\sqrt[95]{360}^{9}$
$\sqrt[95]{360}^{17}$	$\sqrt[95]{360}^{24}$	$\sqrt[95]{360}^{31}$	$\sqrt[95]{360}^{8}$	$\sqrt[95]{360}^{15}$

To check the values, unless we are mathematicians, we will require the use of an advanced scientific calculator or an online scientific calculator, personally I recommend the page desmos.com, on this site you will find very valuable mathematical tools, such as a powerful calculator scientific with algebraic notation.

FIGURE 49

$$\left(\sqrt[95]{360}^{23}\right)\left(\sqrt[95]{360}^{30}\right)\left(\sqrt[95]{360}^{7}\right)\left(\sqrt[95]{360}^{14}\right)\left(\sqrt[95]{360}^{14}\right) = 360$$

(4.158093872) (6.41583847) (1.54297586) (2.380774506) (3.673477591) = 360

Until now we have studied the multiplicative method applied to the primary magic constant, but will this method behave in the same way with the secondary magic constant obtained through additions that we studied in chapter 2??

The answer is NO, this method differs a little bit, since to obtain the value of the magic constant by multiplication, we must in all cases multiply the 4 cells that surround the central position and divide the value of those 4 cells by the center cell.

FIGURE 50

$$\frac{\left(\sqrt[95]{360^{23}}\right)\left(\sqrt[95]{360^{21}}\right)\left(\sqrt[95]{360^{15}}\right)\left(\sqrt[95]{360^{17}}\right)}{\left(\sqrt[95]{360^{19}}\right)}$$

$$\frac{\left(\sqrt[95]{360^{7}}\right)\left(\sqrt[95]{360^{28}}\right)\left(\sqrt[95]{360^{10}}\right)\left(\sqrt[95]{360^{31}}\right)}{\left(\sqrt[95]{360^{19}}\right)}$$

$$\frac{\left(\sqrt[95]{360^{11}}\right)\left(\sqrt[95]{360^{20}}\right)\left(\sqrt[95]{360^{18}}\right)\left(\sqrt[95]{360^{27}}\right)}{\left(\sqrt[95]{360^{19}}\right)}$$

$$\frac{\left(\sqrt[95]{360^{14}}\right)\left(\sqrt[95]{360^{5}}\right)\left(\sqrt[95]{360^{24}}\right)\left(\sqrt[95]{360^{29}}\right)}{\left(\sqrt[95]{360^{19}}\right)}$$

Enter these values into your scientific calculator and you will get the value of 360, which is our multiplicative magic constant for this example over all possible combinations of secondary magic constants..

¿where is the trick?

Possibly by now you have wondered, where I got the formula to fragment the numbers of the cells and then return them to obtain them through multiplications, believe me it is much simpler than you imagine, and I was inspired by it when I studied logarithmic laws in high school, and it goes something like this: What happens if I square a number and then apply the square root to the result? We get the same number as if we hadn't performed any operation, or if we cubed it and then cubed the result, as seen in the following equations.

$$10^2 = 100 \longrightarrow \sqrt{100} = 10$$
$$10^3 = 1000 \longrightarrow \sqrt[3]{1000} = 10$$

Now, what happens when we multiply two equal values raised to different powers? The result is the same as adding the exponents, as seen in the equation.

$$10^2 \times 10^3 = 10^{2+3} = 10^5$$

Applying the upper concept to the rows and columns of a Kumayiko, we can see where the trick is, and it is precisely to raise the value of the magic constant to the power by sums and then obtain the root of the same index, as can be seen in the figure 50 next page.

FIGURA 51

$$\sqrt[95]{360}^{23} \sqrt[95]{360}^{30} \sqrt[95]{360}^{7} \sqrt[95]{360}^{14} \sqrt[95]{360}^{21} \rightarrow \sqrt[95]{360}^{23+30+7+14+21} = \sqrt[95]{360}^{95} = 360$$

$$\sqrt[95]{360}^{29} \sqrt[95]{360}^{11} \sqrt[95]{360}^{13} \sqrt[95]{360}^{20} \sqrt[95]{360}^{22} \rightarrow \sqrt[95]{360}^{29+11+13+20+22} = \sqrt[95]{360}^{95} = 360$$

$$\sqrt[95]{360}^{10} \sqrt[95]{360}^{12} \sqrt[95]{360}^{19} \sqrt[95]{360}^{26} \sqrt[95]{360}^{28} \rightarrow \sqrt[95]{360}^{10+12+19+26+28} = \sqrt[95]{360}^{95} = 360$$

$$\sqrt[95]{360}^{16} \sqrt[95]{360}^{18} \sqrt[95]{360}^{25} \sqrt[95]{360}^{27} \sqrt[95]{360}^{9} \rightarrow \sqrt[95]{360}^{16+18+25+27+9} = \sqrt[95]{360}^{95} = 360$$

$$\sqrt[95]{360}^{17} \sqrt[95]{360}^{24} \sqrt[95]{360}^{31} \sqrt[95]{360}^{8} \sqrt[95]{360}^{15} \rightarrow \sqrt[95]{360}^{17+24+31+8+15} = \sqrt[95]{360}^{95} = 360$$

Now, once this work is finished, I have dedicated a final section with blank exercises so that you can practice what you have learned a little and you can in turn teach and surprise your friends, family or acquaintances by making Kumayikos just as the ambassador of de La Loubère or in the same way as my father taught it to me more than 30 years ago with a pen on a simple napkin..

God bless you dear reader, I am infinitely grateful for the time and money you had invested to buy this book and read these pages to which I have dedicated so many, many years of my life, we will keep in touch!!

Third Part...

Exercises

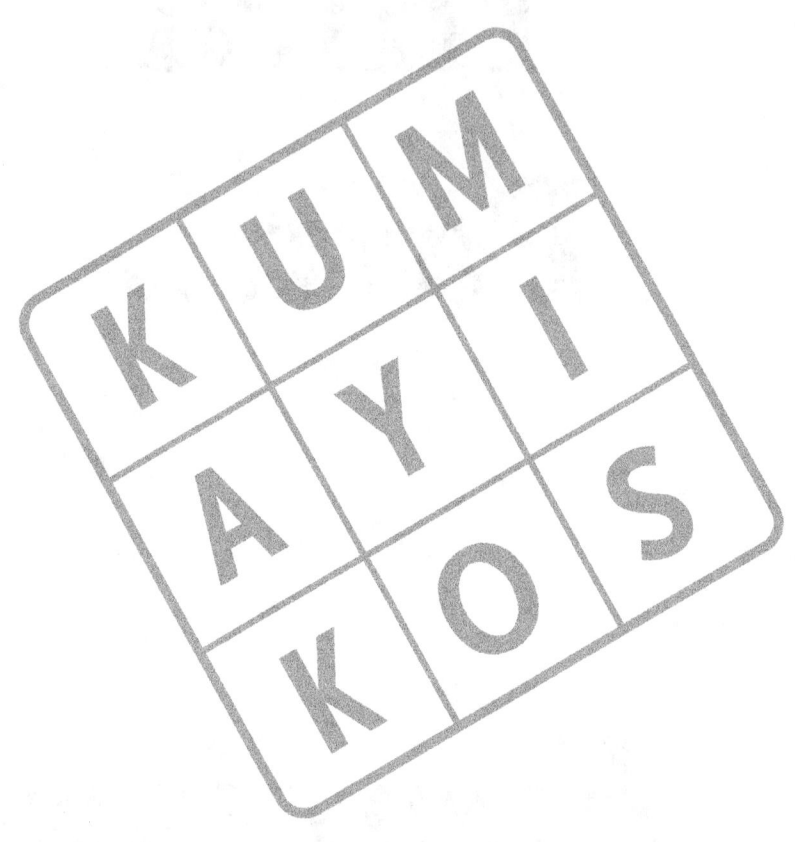

8. Exercises to reinforce

Exercise 1

Complete the following Kumayiko following the Loubère method to obtain a magic constant of 33.

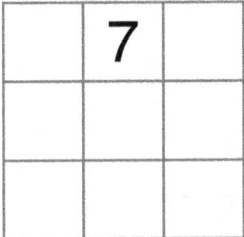

Exercise 2

Following the same method as de La Loubère, complete the following 5x5 Kumayiko to get a magic constant of 85.

Exercise 3

Complete the following 7x7 Kumayiko following the de La Loubère method to obtain a magic constant of 231.

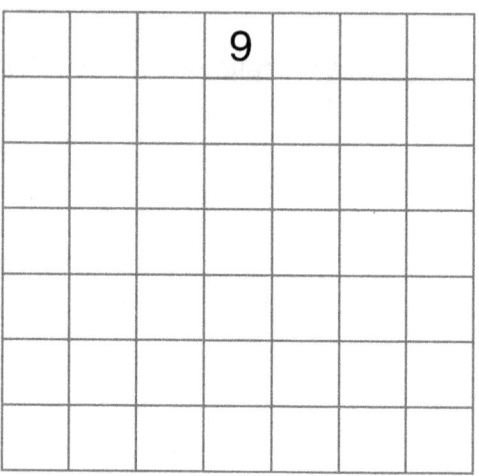

Exercise 4

And now proceed to complete the following 9x9 Kumayiko using the same La Loubère method to obtain a magic constant of 450.

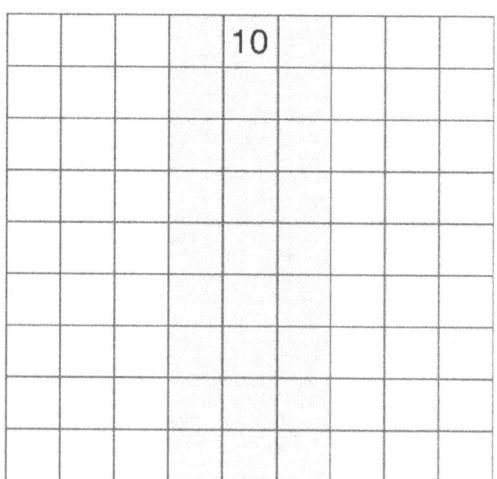

Solution on page 64

Exercise 5

Using the center position method, make a 5x5 Kumayiko whose magic constant is 480.

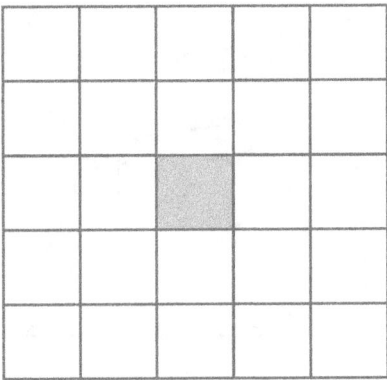

Tip: Instructions on page 25

Exercise 6

Now we are going to get a magic constant of 120 using the same 5X5 Kumayiko.

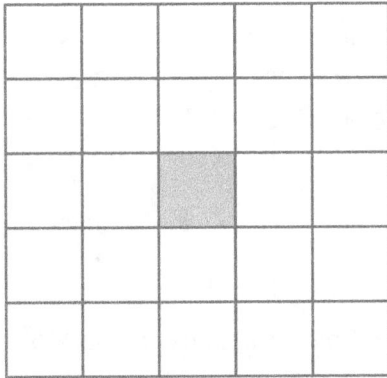

Solution on page 65

Exercise 7

Complete the following 3x3 Kumayiko following the Lu Shu Theorem solving method.

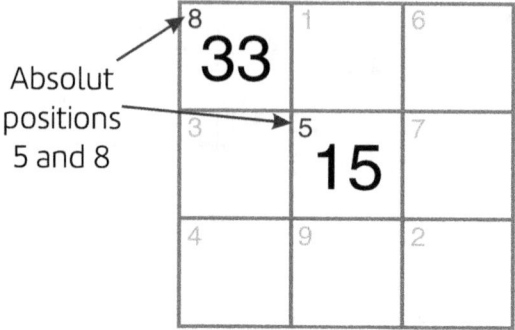

Tip: Instructions on page 25

Exercise 8

Now let's complete the following 3x3 Kumayiko following the Lu Shu Theorem solving method.

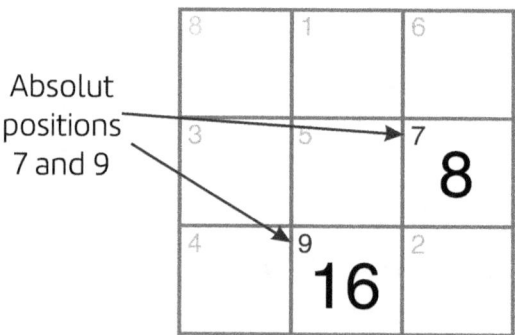

Tip: Instructions on page 25

Solution on page 64

Exercise 9

Now let's make a 5x5 Kumayiko following the method of Lu Shu Theorem in which the numbers 5 and 17 occupy positions 8 and 12.

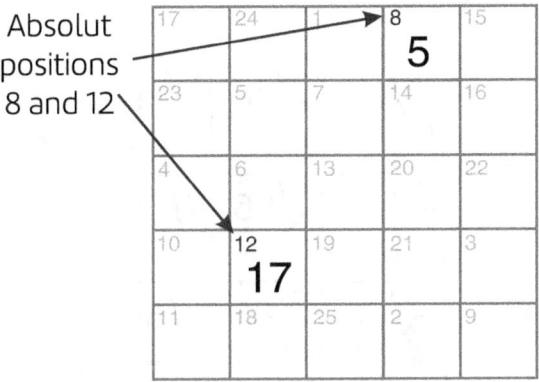

Absolut positions 8 and 12

Tip: Instructions on page 25

Exercise 10

Another Kumayiko more than 5x5 following the same method for Lu Shu theorem.

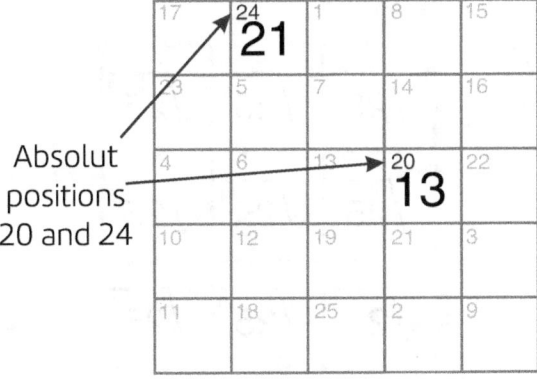

Absolut positions 20 and 24

Solution on page 64

Exercise 11

Put the exponents in each cell to complete the following multiplicative Kumayiko whose primary magic constant for addition is 15 and multiplicative magic constant is 66.

$$\begin{array}{|c|c|c|} \hline \sqrt[15]{66} & \sqrt[15]{66} & \sqrt[15]{66} \\ \hline \sqrt[15]{66} & \sqrt[15]{66} & \sqrt[15]{66} \\ \hline \sqrt[15]{66} & \sqrt[15]{66} & \sqrt[15]{66} \\ \hline \end{array}$$

Tip: Instructions on page 41

Exercise 12

Put the indices on the radicals in each cell to solve the following 3x3 Kumayiko.

$$\begin{array}{|c|c|c|} \hline \sqrt{25}^{17} & \sqrt{25}^{3} & \sqrt{25}^{13} \\ \hline \sqrt{25}^{7} & \sqrt{25}^{11} & \sqrt{25}^{15} \\ \hline \sqrt{25}^{9} & \sqrt{25}^{19} & \sqrt{25}^{5} \\ \hline \end{array}$$

Tip: Instructions on page 41

Solution on page 68

Exercise 13

Put the values (radicals) in each cell to solve the 5x5 multiplicative Kumayiko and get a multiplication constant of 18.

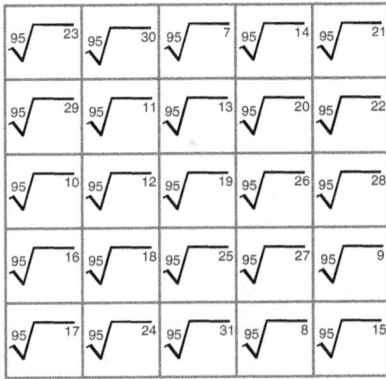

$\sqrt[95]{23}$	$\sqrt[95]{30}$	$\sqrt[95]{7}$	$\sqrt[95]{14}$	$\sqrt[95]{21}$
$\sqrt[95]{29}$	$\sqrt[95]{11}$	$\sqrt[95]{13}$	$\sqrt[95]{20}$	$\sqrt[95]{22}$
$\sqrt[95]{10}$	$\sqrt[95]{12}$	$\sqrt[95]{19}$	$\sqrt[95]{26}$	$\sqrt[95]{28}$
$\sqrt[95]{16}$	$\sqrt[95]{18}$	$\sqrt[95]{25}$	$\sqrt[95]{27}$	$\sqrt[95]{9}$
$\sqrt[95]{17}$	$\sqrt[95]{24}$	$\sqrt[95]{31}$	$\sqrt[95]{8}$	$\sqrt[95]{15}$

Tip: Instructions on page 41

Exercise 14

Put the indices of the radicals and the exponents to complete the following 5x5 multiplicative Kumayiko.

$\sqrt{99}$	$\sqrt{99}$	$\sqrt{99}$	$\sqrt{99}$	$\sqrt{99}$
$\sqrt{99}$	$\sqrt{99}$	$\sqrt{99}$	$\sqrt{99}$	$\sqrt{99}$
$\sqrt{99}$	$\sqrt{99}$	$\sqrt{99}$	$\sqrt{99}$	$\sqrt{99}$
$\sqrt{99}$	$\sqrt{99}$	$\sqrt{99}$	$\sqrt{99}$	$\sqrt{99}$
$\sqrt{99}$	$\sqrt{99}$	$\sqrt{99}$	$\sqrt{99}$	$\sqrt{99}$

Solution on page 69

Fourth Part...

Solutions

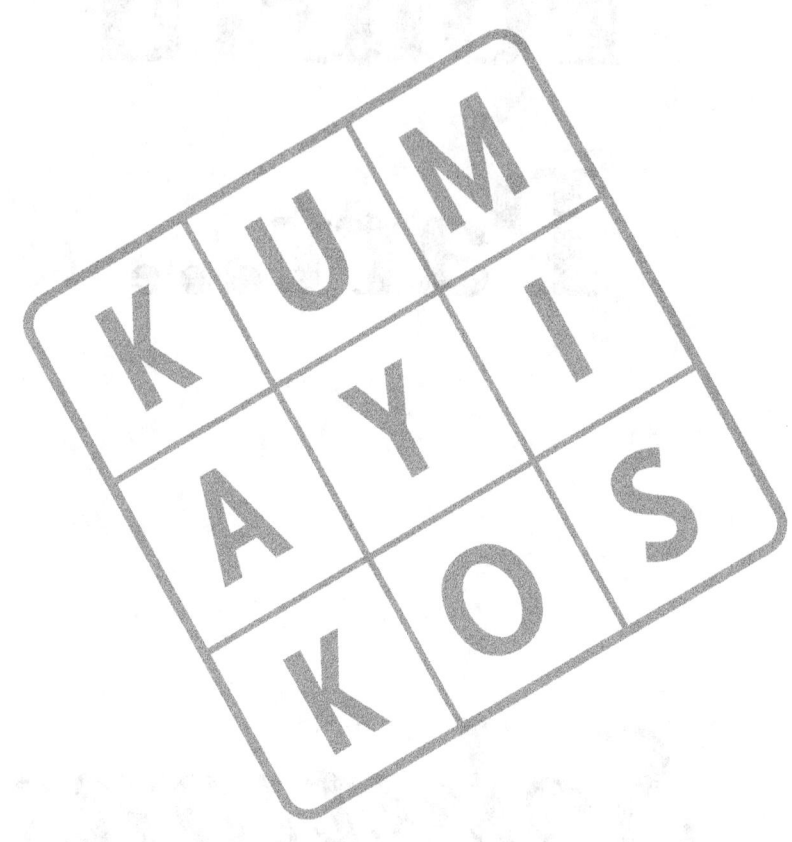

9. Solutions

Solution to exercise 1

Kumayiko broken down in which you can see that the magic constant results 33 in all cases.

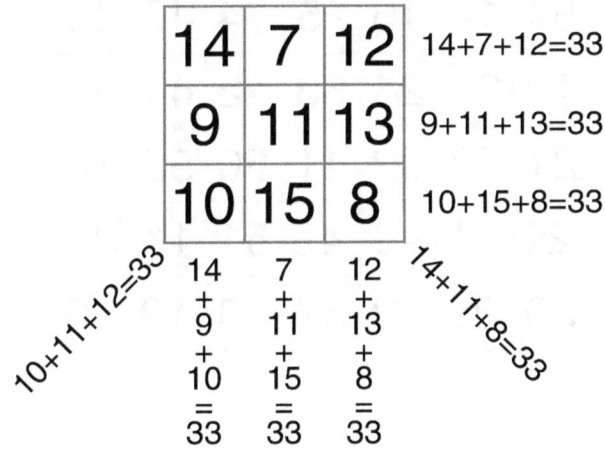

Solution to exercise 2

Kumayiko with magic constant of 85.

21	28	5	12	19
27	9	11	18	20
8	10	17	24	26
14	16	23	25	7
15	22	29	6	13

Solution to exercise 3

Kumayiko with magic constant of 231.

38	47	56	9	18	27	36
46	55	15	17	26	35	37
54	14	16	25	34	43	45
13	22	24	33	42	44	53
21	23	32	41	50	52	12
29	31	40	49	51	11	20
30	39	48	57	10	19	28

Solution to exercise 4

Kumayiko with magic constant of 450.

56	67	78	89	10	21	32	43	54
66	77	88	18	20	31	42	53	55
76	87	17	19	30	41	52	63	65
86	16	27	29	40	51	62	64	75
15	26	28	39	50	61	72	74	85
25	36	38	49	60	71	73	84	14
35	37	48	59	70	81	83	13	24
45	47	58	69	80	82	12	23	34
46	57	68	79	90	11	22	33	44

Solución to exercise 5

Kumayiko with magic constant of 480

100	107	84	91	98
106	88	90	97	99
87	89	96	103	105
93	95	102	104	86
94	101	108	85	92

Solución to exercise 6

Kumayiko with magic constant of 120

28	35	12	19	26
34	16	18	25	27
15	17	24	31	33
21	23	30	32	14
22	29	36	13	20

Solution to exercise 7

By solving this Lu Shu theorem we obtain a magic constant of 45.

Absolut positions 5 and 8

33	-9	21
3	15	27
9	39	-3

Solution to exercise 8

By solving this Lu Shu theorem we get a magic constant of 0

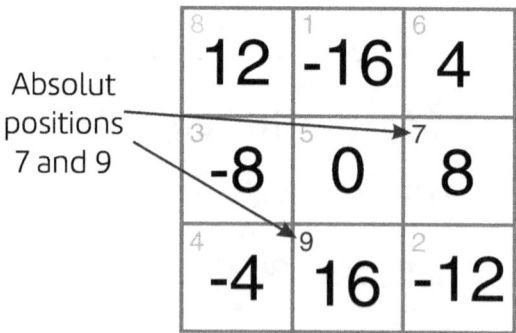

Absolut positions 7 and 9

12	-16	4
-8	0	8
-4	16	-12

Solution to exercise 9

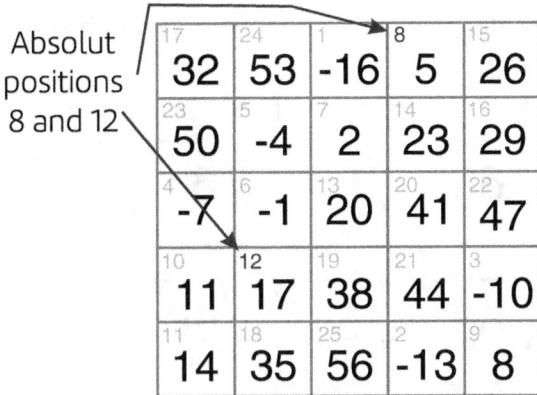

Absolut positions 8 and 12

32	53	-16	5	26
50	-4	2	23	29
-7	-1	20	41	47
11	17	38	44	-10
14	35	56	-13	8

Magic constant of 100

Solution to exercise 10

Absolut positions 20 and 24

7	21	-25	-11	3
19	-17	-13	1	5
-19	-15	-1	13	17
-7	-3	11	15	-21
-5	9	23	-23	-9

Magic constant of −5

Solution to exercise 11

Kumayiko with multiplicative magic constant of 66 through addition magic constant of 15.

$$\begin{array}{|c|c|c|}\hline \sqrt[15]{66^8} & \sqrt[15]{66^1} & \sqrt[15]{66^6} \\\hline \sqrt[15]{66^3} & \sqrt[15]{66^5} & \sqrt[15]{66^7} \\\hline \sqrt[15]{66^4} & \sqrt[15]{66^9} & \sqrt[15]{66^2} \\\hline\end{array}$$

Solution to exercise 12

Kumayiko with multiplicative magic constant of 25 through addition magic constant of 33.

$$\begin{array}{|c|c|c|}\hline \sqrt[33]{25^{17}} & \sqrt[33]{25^3} & \sqrt[33]{25^{13}} \\\hline \sqrt[33]{25^7} & \sqrt[33]{25^{11}} & \sqrt[33]{25^{15}} \\\hline \sqrt[33]{25^9} & \sqrt[33]{25^{19}} & \sqrt[33]{25^5} \\\hline\end{array}$$

Solution to exercise 13

Kumayiko with multiplicative magic constant of 18 through addition magic constant of 95.

$\sqrt[95]{18}^{23}$	$\sqrt[95]{18}^{30}$	$\sqrt[95]{18}^{7}$	$\sqrt[95]{18}^{14}$	$\sqrt[95]{18}^{21}$
$\sqrt[95]{18}^{29}$	$\sqrt[95]{18}^{11}$	$\sqrt[95]{18}^{13}$	$\sqrt[95]{18}^{20}$	$\sqrt[95]{18}^{22}$
$\sqrt[95]{18}^{10}$	$\sqrt[95]{18}^{12}$	$\sqrt[95]{18}^{19}$	$\sqrt[95]{18}^{26}$	$\sqrt[95]{18}^{28}$
$\sqrt[95]{18}^{16}$	$\sqrt[95]{18}^{18}$	$\sqrt[95]{18}^{25}$	$\sqrt[95]{18}^{27}$	$\sqrt[95]{18}^{9}$
$\sqrt[95]{18}^{17}$	$\sqrt[95]{18}^{24}$	$\sqrt[95]{18}^{31}$	$\sqrt[95]{18}^{8}$	$\sqrt[95]{18}^{15}$

Solution to exercise 14

Kumayiko with multiplicative magic constant of 99 through addition magic constant of 95.

$\sqrt[95]{99}^{23}$	$\sqrt[95]{99}^{30}$	$\sqrt[95]{99}^{7}$	$\sqrt[95]{99}^{14}$	$\sqrt[95]{99}^{21}$
$\sqrt[95]{99}^{29}$	$\sqrt[95]{99}^{11}$	$\sqrt[95]{99}^{13}$	$\sqrt[95]{99}^{20}$	$\sqrt[95]{99}^{22}$
$\sqrt[95]{99}^{10}$	$\sqrt[95]{99}^{12}$	$\sqrt[95]{99}^{19}$	$\sqrt[95]{99}^{26}$	$\sqrt[95]{99}^{28}$
$\sqrt[95]{99}^{16}$	$\sqrt[95]{99}^{18}$	$\sqrt[95]{99}^{25}$	$\sqrt[95]{99}^{27}$	$\sqrt[95]{99}^{9}$
$\sqrt[95]{99}^{17}$	$\sqrt[95]{99}^{24}$	$\sqrt[95]{99}^{31}$	$\sqrt[95]{99}^{8}$	$\sqrt[95]{99}^{15}$

Dear reader, for the elaboration of this book I invested a lot of time of my life, but if you detect any error in the content, I would be very grateful if you would notify me through an email indicating the page and the error to correct it in future edits.

electronic mail: **kumayikos@gmail.com**

www.ingramcontent.com/pod-product-compliance
Lightning Source LLC
Chambersburg PA
CBHW050253220526
45465CB00002B/657